Small Beginnings

BOOK 1 IN THE SERIES
EVER EXPANDING UNIVERSE

Marilyn D Corner

Copyright © 2017 by Marilyn D Corner.
All Rights Reserved.

No part of this book may be reproduced in any form or by any electronic or mechanical means including information storage and retrieval systems, without permission in writing from the author. The only exception is by a reviewer, who may quote short excerpts in a review.

First Printing: November 2017

Cover photo: Hubble Space Telescope photo of the planetary nebula NGC 2818, one of few planetary nebulae in the Milky Way residing inside a star cluster. A planetary nebula is an expanding shell of gas surrounding a dying star

Small Beginnings / Marilyn D Corner. —1st ed.
ISBN 978-1979510530

Contents

Preface 1

Time Traveller 3

Space Traveller 9

Finding Mum & Dad 13

Evolution 15

Human Evolution 21

Genetics 33

Stardust 39

Galaxies 43

Solar System 47

The Sun 57

The Moon 61

The Earth 67

The Future 77

Quiz 89

Timeline of Universe 91

Glossary 97

Answer to quiz questions 105

FOR THE LOVE OF MY LIFE

Thank you for everything

There are places out there, billions of places out there, that we know nothing about.

And the fact that we know nothing about them excites me, and I want to go out and find out about them.

And that's what science is.

—BRIAN COX, BRITISH PHYSICIST

Preface

Science is a way of knowing about the natural world. The methods and disciplines of science provide the only valid route to gaining true knowledge and it is so obviously the way forward in the 3rd millennium.

Science has made such huge advances in recent years, on a whole range of subjects, that we now know so much more than we did even 10 years ago, let alone 100 or 1,000 or 2,000 years ago.

The problem is that it can be really difficult to keep pace, as the sheer volume of information now being produced means that we are probably learning less and less about more and more. And let's face it, there are many levels to science, some of which may go way over our heads.

However, let's not be intimidated because there is enough of interest for everyone at every level.

Did you know, for example, that you were made entirely of stardust?

You didn't?

Well, read on!

CHAPTER 1

Time Traveller

This story starts, as you might have guessed, with a bang.

Actually, with a "Big Bang" – the one that created our universe as we know it, something in the region of 14 billion years ago.

Now at this point in time, it really doesn't matter or concern us too much as to what caused the Big Bang or what, if anything, came before it but you can be sure that science will continue to strive to find this out, because that is what science does.

In fact, recent thinking now is to the effect that there could be more than one universe, possibly many universes, and that our universe is just one event that is happening over and over again in some "multiverse".

That has some appealing logic to it as it does away with the problem of explaining how our universe came to be started from nothing or, even less likely, was created by some supernatural being or other.

Even less than 100 years ago, our universe was thought to be static and eternal. It was only in 1929 that American astronomer Edwin Hubble discovered that other galaxies were actually travelling away from us. This showed that the universe was expanding and was the first real evidence of the beginning of the universe, the Big Bang.

The Big Bang was not an explosion of matter moving outward to fill an existing empty universe. Instead, space itself expands with time everywhere. In other words, the Big Bang is not an explosion *in space*, but rather an expansion *of space*.

What we know so far is that this Big Bang consisted of pure energy, originally smaller than an atom, and this expanded into a huge universe in something less than a second. Since that time, the universe has continued to expand at an incredible rate.

We do not know exactly where we sit within the universe but our "field of view" is currently said to be of the order of 150 billion light years wide - with one light year being the equivalent of 6 trillion miles. It's enormous.

Science has now managed to recreate the history of the Big Bang to within less that a millionth of a second of the explosion and hopes to improve on that experimentally using the Large Hadron Collider in Switzerland.

Everything in the universe was brought into being from matter created in the first seconds of the Big Bang. Most of the matter in the cosmos at this stage was dark matter with the little remaining ordinary matter comprised largely of neutral hydrogen and helium.

Some 380 million years later, when the universe had started to cool to a temperature something in the order of 18 million degrees Fahrenheit, the energy started to transform into sub-atomic particles and later into simple atoms. The first atom to form was hydrogen and when hydrogen atoms collided with each other another element, helium, was formed. This, in time, resulted in the formation of electrically charged, neutral transparent gas.

The initial flash of light created is detectable today as cosmic microwave background radiation, the afterglow of the Big Bang's heat. The Cosmic Microwave Background is radiation that fills the universe and can be discovered in every direction. It represents the earliest radiation that can be detected.

However, after this point, the universe fell back into darkness and it was not until about 400 million years after the Big Bang that the universe began to emerge from these dark beginnings, when these dark, dense, clouds of gas collapsed, that the first stars and galaxies were formed.

From this moment on, what was a very dark universe suddenly began to see light and it is the light from these stars and galaxies that we see in the night sky now.

Remember that we are not looking at the stars directly, but only at the light emitted from them all that long time ago. It is a fact that the stars that we "see" now could well have died out millions of years ago, but the light is still travelling across the universe towards us.

For example, the light currently reaching us from the North Star was emitted as long ago as 1583, at a time when Elizabeth 1 was on the throne of England and only a year or two after Francis Drake returned from his voyage around the World.

We are literally staring into the past.

During the huge expansion of space that followed, the expansion of the universe gradually slowed down as gravity pulled the matter together.

Then, about 6 or 7 billion years after the Big Bang, this mysterious force that we now refer to as to dark energy began speeding up the expansion of the universe again, something that continues to this day.

The stars, like our very own Sun, can be described as nuclear cauldrons and their enormous temperatures and pressures led to the fusion of other elements such as carbon, nitrogen, oxygen and iron.

It is estimated that one star dies every second. When these bigger stars eventually become exhausted and die they explode, something called a supernova, and they throw out all of these newly forged elements into their galaxy.

Solar winds can then blow them across space to other galaxies.

When enough of these elements gathered together, under the force of gravity, to form a star some 4.6 billion years ago, our own Sun was born and the leftover material was used to form the planets of our Solar System including the Earth.

Now the universe can be defined as everything that exists, everything that has existed, and everything that will exist.

That means that you undoubtedly existed all those years ago, you have continued to exist all this time and you exist now.

Obviously, you did not exist in human form all those years ago but you did exist in one form or another.

You have been around for 13.8 billion years and you are still here.

So, you are right up there with Dr. Who!

You are some time traveller!

> *"Any man who can hitch the length and breadth of the galaxy, rough it, slum it, struggle against terrible odds, win through, and still knows where his towel is, is clearly a man to be reckoned with."*
>
> *— Douglas Adams, The Hitchhiker's Guide to the Galaxy*

CHAPTER 2

Space Traveller

The fact that you presently exist in an intelligent human form means that this is your moment in time, and time to make the most of your existence. Whatever you do, you must not waste the opportunity that you now have and one that will never occur again!

Although you have been around for all of these 13.8 billion years, you have never had the opportunity to become aware of the fact until now. It is only now, in your human form, that you have the opportunity to be aware of your existence.

Remember, you are one of the very lucky ones. The unlucky ones are the ones who will never be born, never have the chance to have lived at all. Think about it.

The universe is known to contain trillions of galaxies and, in each galaxy, is to be found billions of stars and planets & moons. It is said that there are as many galaxies in our universe as there are grains of sand on Earth.

Despite the fact that there are billions and trillions of such objects, known as ordinary or baryonic matter, they still take up only approx. 4% of the universe. The remainder of the universe is "Space" and this is made up, not of emptiness, but largely of a lethal, radiation filled, hard vacuum, also containing little understood dark matter, dark energy, black holes, etc. And this Space actually takes up some 96% of the universe

Our Sun is a star and together with the Earth and other planets make up our Solar System and this sits in our galaxy known as the Milky Way.

Given that there are trillions of such galaxies, that makes your achievement of finding your way to the only galaxy, then to the only solar system and then to the only planet in the universe, Earth, that we know of that is hospitable to human life, even more incredible.

Even where you live on Earth, 70% of the planet is covered by oceans of water and of the remaining 30% of land, two thirds is either too hot or too cold. You have also just happened to find the 10% area on Earth that has just the right temperature.

You have just happened to have found your way to the ideal goldilocks location.

You have proved yourself to be pretty capable at orienteering.

So, you are also an excellent space traveller.

*"The universe is a pretty big place.
If it's just us,
Seems like an awful waste of space."*

Carl Sagan, American astronomer, Cosmologist,
Astrophysicist, Astrobiologist

CHAPTER 3

Finding Mum & Dad

But your achievement doesn't end there.

For you to be born you still have the enormous task of finding the exact two people on Earth capable of giving you exactly the right genes. Half from your Mum and half from your Dad.

And with the Earth's population in the order of 8 billion people, that is some needle in a haystack. But you did it. Amazing.

And there is another fact that is very relevant to your journey – the fact that of all the species on Earth, 99% have already died out. Even the human race at one time is known to have been reduced to no more than a few thousand people.

It is generally accepted that over time there have been five major mass extinctions on Earth and geologists have now fixed on the precise event that set off Earth's most devastating mass extinction, approximately 252 million years ago, which killed off 90 percent of all marine organisms and 75 percent of all life on land.

At this time, a huge pulse of magma rose up through the Earth, some of which molten liquid stopped short of erupting onto the surface and instead spread out beneath the Earth's shallow crust, creating a vast network of rock stretching across almost one million square miles.

This heated the surrounding carbon-rich sediments and rapidly released into the atmosphere a tremendous volume of carbon dioxide, methane, and other greenhouse gases.

In another mass extinction 66 million years ago, caused by a massive comet or asteroid impact, it is estimated that 75% or more of all species on Earth vanished, including the dinosaurs.

Five mass extinctions plus many more smaller extinctions and yet you have survived.

Is there no end to your talents!

And there's still more...

CHAPTER 4

Evolution

So, how did your parents get here?

We know that living things have changed over time, because we can see their fossil remains in the rocks that make up the Earth's crust. We know that the animals and plants of today are different from those of long ago and the further we go back in time, the more different the fossils are.

Recent research has shed fresh light on how animals first appeared on the Earth. This "revolution of ecosystems" began to happen about 700 million years ago when the planet became what is referred to as 'Snowball Earth' as ice covered virtually its entire surface.

Giant glaciers ground down huge mountains to powder, releasing life-giving nutrients such as phosphates, and then, during an "extreme global heating event", the Snowball melted and rivers washed the nutrients into the ocean.

The newly temperate climate and the sudden influx of food created the perfect conditions for the first complex life – algae – that are the ancestors of us all.

In 1859, Charles Darwin published his book, "On the Origin of the Species", on the theory of evolution by means of natural selection.

Evolution is a scientific theory used by biologists. It explains how living things change over a long period of time, and how they have come to be the way they are.

Natural selection is the naturalistic equivalent to domestic breeding. Over the centuries, human breeders have produced dramatic changes in domestic animal populations by selecting individuals to breed. Breeders eliminate undesirable traits gradually over time.

As random genetic alterations, or mutations, occur within an organism's genetic code, the beneficial mutations are preserved because they aid survival -- a process known as "natural selection." These beneficial mutations are then passed on to the next generation.

Over time, beneficial mutations accumulate and the result is an entirely different organism - not just a variation of the original, but an entirely different creature.

Vestigial organs are organs of the body which are smaller and simpler than those in related species. These organs have lost, or almost lost, their original function.

> ***It is not the strongest of the species that survives, nor the most intelligent, but the one most responsive to change.***
>
> *Charles Darwin*

This is further evidence of evolution, since they only make sense if evolution has occurred. They were one of the puzzles of pre-Darwinian natural history biologists.

The puzzle was solved once it was realized that these organs were once working adaptions in the ancestors of present-day animals.

They occur in animals and plants which have changed their style of life from their ancestors. Thus, snakes lost their legs as their system of movement changed. But one type of snake – the boas – still have vestigial rear legs and pelvis.

And the same applies to humans where the appendix is another example. Originally much larger, it was used to produce cellulose to break down plant cell walls. Whilst plants remain the main diet of apes, they are no longer a main part of man's diet. Cellulose cannot now be digested by man.

Another example is the fact that all mammals have a laryngeal nerve that makes a pointless detour down and up the neck again. In the fish that we evolved from this was the most efficient route, but our blue print comes from them and not from a designer, so we make do with what we have.

In humans, the strong grip of a baby is another example. It is a vestigial reflex, a remnant of the past when pre-human babies clung to their mothers' hair as the mothers swung through the trees. This is borne out by the babies' feet, which curl up when it is sitting down (primate babies grip with the feet as well).

Some biologists say that evolution has happened when a trait that is caused by genetics becomes more or less common in a group of organisms. Others call it evolution when a totally new species appears.

Changes can happen quickly in the smaller, simpler organisms. For example, many bacteria that cause disease can no longer be killed with some of the antibiotic medicines.

These medicines have only been in use about eighty years, and at first worked extremely well but the bacteria have evolved so that they are no longer affected by antibiotics anymore.

The drugs killed off all the bacteria except a few which had some resistance. These few resistant bacteria produced the next generation.

Evolution only concerns the traits which can be inherited, wholly or partly. The hereditary traits are passed on from one generation to the next through the genes.

A person's genes contain all the traits which they inherit from their parents. The accidents of life are not passed on. Similarly, where we have a situation like our eyesight getting progressively worse as we age, this has never been improved by natural selection because as we reach these ages we are no longer giving birth to children.

Evolution is one of the most successful theories in science. People have discovered it to be useful for different kinds of research. None of the other suggestions explain things, such as fossil records, as well. So, for almost all scientists, evolution is not in doubt.

> *The theory of evolution, like the theory of gravity, is a scientific fact.*
>
> Neil deGrasse Tyson, American Astrophysicist

Another fact, too, is that the Earth teems with a staggering variety of creatures, including 9,000 kinds of birds, 28,000 types of fish, and more than 400,000 species of beetles.

The source of life's endless forms was a profound mystery until Charles Darwin brought forth his revolutionary idea of natural selection.

Thomas Henry Huxley later applied Darwin's ideas to humans, using paleontology and comparative anatomy, to provide strong evidence that humans and apes shared a common ancestry.

But Darwin didn't know any of the mechanics of the evolution process. He didn't know what was happening inside a creature's body to make it change.

Modern science is now providing those answers and the answer lies in the DNA of cells.

> *The universe we observe has precisely the properties we should expect if there is, at bottom, no design, no purpose, no evil, no good, nothing but blind, pitiless indifference.*
>
> Charles Darwin, English Naturalist, geologist and biologist

CHAPTER 5

Human Evolution

Humans are primates. Physical and genetic similarities show that the modern human species, Homo sapiens, has a very close relationship to another group of primate species, the apes. However, to firstly dispose of a common myth, mankind is not descended from chimps or even from apes for that matter.

The original apes, (Hominoidea) are native to Africa and Southeast Asia

Within this main apes superfamily, the Hominidae family diverged from the gibbon family some 15–20 million years ago and African great apes diverged from orangutans about 14 million years ago. The Hominini tribe (humans, *Australopithecines* and other extinct bipeds and chimpanzee) parted from gorillas between 8 and 9 million years ago and, in turn, the subtribes *Hominina* (humans and biped ancestors) and chimps separated about 5.6 to 7.5 million years ago.

The word *hominidae* describes the total member species of the human family that have lived since the last common ancestor of both man and the apes. We are, as yet, unaware of just who that common ancestor might be. A *hominid* is an individual species within that family.

Because we did have a common ancestor it follows that man and the apes did look very similar all that time ago and it is mankind that has changed in appearance quite rapidly ever since.

There are thousands of fossils that document the developing human-like species but the problem in many cases is that only one or two bones are found and never a whole skeleton. There is then the difficulty of aging these fossils given that DNA testing is not feasible. Most dating is therefore carried out based on the soil strata where the fossils are found as in many cases there will be evidence of other creatures with a known time-line.

The finds are then classified by age into different species names with the most recent species being *Homo sapiens*, (wise man), to which we all belong.

The current line between *Homo* and previous non-*Homo* species is drawn about 2.2 million years ago, with the *Australopithecus* ranging back from there to about 4.2 million years ago and prior to that we have *Ardipithecus* ranging back to some 6 million years.

So, on what is known at present, *Ardipithecus* is the oldest species of modern man and the nearest to our common ape ancestry.

The evolutionary changes taking place over this time relate firstly to bi-pedal activity from when man first walked upright, to changes in the shape of the skull to accommodate an increasing brain size. The original sloping forehead, the heavy sagittal crest running backwards at the top of the skull, the heavy ridge along the line of the eyebrows and the changes in the shape of the jaw are all features of such changes.

Ardipithecus kadabba

Ardipithecus kadabba, (oldest ancestor), lived in Ethiopia between 5.2 and 5.8 million years ago and was discovered in 1997. Originally thought to be a sub-species of *Ardipithecus ramidus*, it was only in 2002 that further finds established that this was a new species. *Ardipithecus kadabba* was bipedal (walked upright), probably had a body and brain size similar to a modern chimpanzee, and had canines that resemble those in later hominins but that still project beyond the tooth row. This early human species is only known in the fossil record by a few post-cranial bones and sets of teeth. One bone from the large toe has a broad, robust appearance, suggesting its use in bipedal push-off.

Ardipithecus ramidus 'Ardi'

Before that, scientists formally announced and published the findings of a partial female skeleton, dating back to 4.4 million years and nicknamed 'Ardi', originally discovered in Ethiopia between 1992 and 1994.

At the time of this discovery, the genus Australopithecus was already scientifically well established, so the genus name Ardipithecus ramidus was used to distinguish this new and older genus from Australopithecus.

The word 'ramid' means 'root' and refers to the closeness of this new species to the roots of humanity, while 'Ardi' means 'ground' or 'floor'.

The skeleton reflects a human-African ape common ancestor that was not chimpanzee-like with both tree-climbing and bi-pedal features. There was also evidence to show that 'Ardi' lived in a wooded environment as opposed to open plains.

Australopithecus anamensis

In 1994, numerous teeth and fragments of bone were found at a site in Kenya. It was determined that the fossils were those of a very primitive hominin and they named a new species called A*ustralopithecus anamensis* ('anam' means 'lake'). This species has a combination of traits found in both apes and humans.

Researchers have since found many other similar fossils at nearby sites, all of which date between about 4.2 million and 3.9 million years old.

Australopithecus afarensis 'Lucy'

Australopithecus afarensis is one of the longest-lived and best-known early human species—nicknamed Lucy – after the discovery in 1974.

Remains have now been uncovered from more than 300 individuals dating back from 2.95 to 3.85 million years ago in Eastern Africa. This species survived for more than 900,000 years, which is over four times as long as our own species has been around.

Displaying both ape and human characteristics, males were 5 foot tall on average and females, 3 feet 6 inches on average.

They also had small canine teeth like all other early humans, and a body that stood on two legs and regularly walked upright. Their adaptations for living both in the trees and on the ground helped them survive for almost a million years as climate and environments changed.

"Lucy" acquired her name from the song "Lucy in the Sky with Diamonds" by The Beatles, which was played loudly and repeatedly in the expedition camp all evening to celebrate the find, after the excavation team's first day of work on the recovery site. After public announcement of the discovery, Lucy captured much public interest, becoming a household name at the time.

Australopithecus africansis 'The Taung child'

Autralopithecus africansis lived 2.1 to 3.3 million years ago in Southern Africa.

Anatomically similar to Australopithecus anemensis with a larger brain and smaller teeth. Its anatomy indicates that it was adapted for walking upright but still retained climbing abilities.

The Taung child, found in 1924, was the first to establish that early fossil humans occurred in Africa.

Australopithecus robustus 'Eurydyce'

Australopithecus robustus or *Paranthropus robustus* was the first discovery made in 1938 of a "robust" species of hominin found in southern Africa living 1.2 to 1.8 million years ago. "Robust" refers, not to the body shape but rather to the teeth and jaw.

It had large jaws and jaw muscles with the accompanying sagittal crest, and post-canine teeth that were adapted to serve in the dry environment they lived in.

The skull of "Eurydice", possibly a female was later discovered in 1994 also in Southern Africa and dating back to 2.3 million years ago.

Homo habilis 'Jonny's child'

Homo habilis lived in Tanzania in Africa some 1.4 to 2.4 million years ago.

Discovered in 1960 and called 'Jonny's child' after the person who found the fossils. When it was realised that this specimen was markedly different from Australopithecus it was named Homo habilis, meaning 'Handy man', because of the large amount of stone tools found at the site.

Homo erectus Trinil 2, Turkana boy, Java man and Peking man

Homo erectus living between 143,000 and 1.89 million years ago, lived in the north, east and south of Africa, as well as in Asia where they have been discovered in Georgia in West Asia, and in China and Indonesia in East Asia. There is currently no evidence of them having inhabited Europe.

They are the oldest known early humans to have possessed modern human-like body proportions with relatively elongated legs and shorter arms compared to the size of the torso. This suggests that they were adapted to a life on the ground, with a loss of their earlier tree-climbing adaptations. They ranged in height from 4 foot 9 inches to 6 foot 1 inch tall.

The first specimen discovered was the skull cap of an individual known as 'Trinil 2' and found in Indonesia in 1891 and later dated to between 700.000 years ago and one million years ago.

A later discovery in 1984 in Kenya, Eastern Africa, of a 40% complete skeleton of a young boy aged 8/9 years was discovered, known as the 'Turkana Boy', dating around 1.6 million years ago. Originally classified as a different species with the name, *Homo Ergaster*, there is still disagreement as to whether this is the ancestor of *Homo Erectus* or is simply of the same species.

Early fossil discoveries from Java in the 1890's, 'Java man', has also been dated about 1.5 million years ago with the further discovery in China in 1927 of 'Peking man'.

Homo erectus is generally believed to have been the first species to have travelled out of Africa, and was possibly the longest living early human species.

Homo sapiens (archaic), Homo heidelbergensis

Neanderthals, Denisovans, and modern humans are all considered to have descended from *Homo heidelbergensis*. Discovered In 1908 near Heidelberg, Germany and dating back about 200,000 to 700,000 years ago this species probably lived in Asia and Africa as well. Males averaged 5 foot 9 inches in height with females averaging 5 foot 2 inches.

They provide the bridge between *Homo erectus* and *Homo sapiens sapiens* during the period 200,000 to 700,000 years ago. Many skulls have been found with features intermediate between the two. Brain averaged about 1200cc and speech was indicated. Skulls are more rounded and with smaller features. Molars and brow ridges are smaller. The skeleton shows a stronger build than modern human but was well proportioned.

This early human species had a very large eyebrow ridge, and a larger braincase and flatter face than older early human species. It was the first early human species to live in colder climates; their short, wide bodies were likely an adaptation to conserving heat. It was the first early human species to routinely hunt large animals. This early human also broke new ground; it was the first species to build shelters, creating simple dwellings out of wood and rock.

Homo Sapiens neandertalensis

Homo sapiens neandertalensis lived in Europe and the Mideast between 400,000 and 40,000 years ago when they became extinct. Neandertals coexisted with *Homo sapiens (archaic)* and early *Homo sapiens sapiens*. It does appear that there was some limited inter-breeding but recent DNA studies have indicated that the *Homo neandertalensis* was an entirely different species and did not merge into the *Homo sapiens sapiens* gene pool.

Fossil remains were discovered in 1829 at Engis, Belgium, and in 1848 at Forbes Quarry, Gibraltar but it wasn't until a third find in Germany in 1856 that they were properly identified as neanderthals.

In 1864, it became the first fossil hominin species to be named after the find in the Neander Valley in Germany, (tal = valley).

Homo sapiens sapiens

Modern humans are the subspecies *Homo sapiens sapiens.*

The species that you and all other living human beings on this planet belong to is Homo sapiens. Anatomically, modern humans can generally be characterized by the lighter build of their skeletons compared to earlier humans. Modern humans have very large brains, which vary in size from population to population and between males and females, but the average size is approximately 1300 cubic centimeters. Housing this big brain involved the reorganization of the skull into what is thought of as "modern" -- a thin-walled, high vaulted skull with a flat and near vertical forehead.

Modern human faces also show much less (if any) of the heavy brow ridges and position of the jaw of other early humans. Our jaws are also less heavily developed, with smaller teeth. Scientists sometimes use the term "anatomically modern Homo sapiens" to refer to members of our own species who lived during prehistoric times.

The facial characteristics of modern man are about 100,000 years old. The faces of earlier hominid were much more apelike.

In 2017, fossils found in Morocco were re-examined and the dating suggests that *Homo sapiens* may have evolved as early as 315,000 years ago in Africa, and other evidence suggests that *Homo sapiens* may have migrated from Africa as early as 270,000 years ago.

Prior to this one recent find it was thought that Homo Sapiens only evolved about 200,000 years ago.

Denisovans

In 2008, whilst digging in a cave in southern Siberia a 40,000-year-old adult tooth was found and a well preserved fossilized finger bone that had belonged to a young girl who was between five and seven years old when she died.

DNA extracted from the finger bone showed the girl was closely related to Neanderthals, yet distinct enough to merit classification as a new species of archaic humans, which scientists named "Denisovan" after the cave where the finger bone was found. The Denisovan genome also suggests the young girl had brown hair, eyes, and skin.

*****

In October 2017, it was reported that a 9.7-million-year-old discovery of fossilised human teeth in Germany is already leading experts to re-examine the birthplace of mankind, previously thought to be in Africa.

The teeth are not of any species that has previously been found in Asia or Europe, but they have the appearance of those that belonged to early hominin skeletons of Lucy,

(Australopithecus afarensis) and Ardi, (Ardipithecus ramidus), that had been found in Ethiopia.

The teeth were found next to the remains of an extinct genus of horse, that helped date the teeth but these new teeth are at least 4 million years older than the African skeletons.

This find, together with the recent Morocco find, may well produce a need for a complete re-thinking of the birthplace of mankind which was hitherto thought to have been concentrated in regions of Eastern and Southern Africa .

CHAPTER 6

Genetics

Genetics is the study of genes, genetic variation, and heredity in living organisms.

Cells are the smallest independent parts of organisms: the human body contains about 100 trillion cells, while very small organisms like bacteria are just one single cell

There is a simple division of labor in cells—genes give instructions and proteins, the elements, carry out these instructions, tasks like building a new copy of a cell, or repairing damage.

The DNA is a long molecule that looks like a twisted ladder. It is the perfect system for storing vast amounts of information that is necessary for building all kinds of creatures. DNA exists in the cells of every living thing on Earth.

DNA contains vital information that gets passed on to each successive generation. It coordinates the making of itself as well as other molecules (proteins). If it is changed even slightly, serious consequences may result. If it is destroyed beyond repair, the cell dies.

DNA can change. One way it can change is when a baby is born it receives half of its genes from each parent. It can also change through mutation. It is mutation that generates variation. Differences between species.

One example of a change through mutation is when one person developed blue eyes instead of the usual brown eyes and this happened only about 6,000 to 10,000 years ago. This is thought to have been at a time when humans migrated out of Africa into Europe because blue eyes were only to be found in Europe.

It was originally thought to be due to mutation of a gene, OCA2, that determined how much brown pigment our eyes have. It is now known, however, that it is not the gene OCA2 itself that is responsible but rather another gene, HERC2, that switches off the brown pigment in OCA2, revealing the underlying blue.

It is believed that this could have had something to do with cross-breeding with another species of humans – possibly the Neanderthals, but there is no direct evidence to support this. It seems probable that all blue-eyed humans are descended from the same ancestor and, again, it is hypothesized that when it happened, blue eyes immediately became very desirable and increased the population accordingly.

Both OCA2 and HERC2 are to be found in chromosome 15 and are also influential in freckling, hair and skin tone. Most babies of European descent have light colored eyes at birth and this changes when about one year old due to increase in melanin at that time.

The nucleotide forms the rungs of the DNA ladder and are the repeating units in DNA. There are four types of nucleotides, two on each rung, known by the letters, A, T, G and C and it is the sequence of these nucleotides that carries information. An A nucleotide must go opposite a T nucleotide, and a G opposite a C.

A chromosome is a package for carrying DNA plus its associated proteins in the cells. They contain a single long piece of DNA that is wound up and bunched together into a compact structure. Different species of plants and animals have different numbers and sizes of chromosomes.

A gene is a segment of DNA that codes for one protein. The function of genes is to provide the information needed to make molecules, called proteins, in cells.

Genes are like sentences made of the "letters" of the nucleotide alphabet, with which they direct the physical development and behavior of an organism.

Genes are like a recipe or instruction manual, providing information that an organism needs so it can build or do something – like making an eye or a leg or repairing a wound.

A genome is the complete set of genes in a particular organism.

Now, if the genome was a book it would contain 23 chapters or chromosomes. Each chapter containing between 48 to 250 million letters, without spaces, a total of 3.2 billion letters and each book would fit into a cell nucleus the size of a pinpoint.

The Human Genome Project was started in 1990 with the goal of sequencing and identifying all three billion chemical units in the human genetic instruction set, finding the genetic roots of disease and then developing treatments.

It is anticipated that detailed knowledge of the human genome will provide new avenues for advances in medicine and biotechnology

The DNA of animals was also to be compared and it was expected humans would have far more. In fact, that was not the case. When the results were published in 2003 it was established that humans had 23,000 genes – the same number as a chicken.

> *We are just an advanced breed of monkeys on a minor planet of a very average star. But we can understand the Universe. That makes us something very special.*
>
> *— Stephen Hawking, British Cosmologist, Theoretical physicist*

Human and chimpanzee chromosomes are also very similar. The primary difference is that humans have one fewer pair of chromosomes than do other great apes. Humans have 23 pairs of chromosomes and other great apes have 24 pairs of chromosomes.

Many key genes in humans were also found to be identical to other animals, so there had to be something more happening. It has since been shown that it is not what genes you have but how you use them that is critical and it now appears that the platform for diversity is in the embryos.

Each of a person's cells holds a copy of all their genes. Although all body cells in an organism contain the same genes, many genes in a particular cell may be 'switched off' (be made inactive) as the cell only produces the specific proteins it requires.

In an embryo, cells are unspecialised up to the eight-cell stage, with all genes in the cells switched on. Cell specialisation will occur as the embryo develops, with different genes switching off in different cells.

CHAPTER 7

Stardust

And, so to the stars.

A star is a luminous sphere of plasma held together by its own gravity. The nearest star to Earth is our Sun.

Many other stars and galaxies are visible to the naked eye from Earth during the night, appearing as a multitude of fixed luminous points in the sky due to their immense distance from the Earth.

In the universe, one star (i.e. a sun) blows up and dies every second. Our sun is forecast to explode and die in about another 5 billion years but, before that, it will gradually warm up until life could become uninhabitable on Earth as early as about 1.5 billion years' time.

Approximately 275 million new stars are born every day - which shows just how immense the universe actually is.

As we have seen, the original chemical elements of the early universe are thought to have been limited to hydrogen and helium. All other atoms were later formed by nuclear reactions within the cores of these ancient stars.

When some larger stars eventually explode, known as a supernova, their many elements are spread far and wide throughout space. There are some 94 naturally occurring elements on Earth with another 24 synthetic elements.

The solar system, including the life upon planet Earth, is made of this star debris.

Everything we are and everything in the universe and on Earth originated from stardust, and it is continually passing through us even today.

It shows that we are very much part of the universe, rebuilding our bodies over and over again in our lifetimes.

All the material in our bodies originates from that residual stardust, possibly from many different stars and galaxies.

It also finds its way into plants, and from there into the nutrients that we need for living. It can be seen that our bodies are therefore being constantly replenished.

In August 2014, scientists announced the collection of possible interstellar dust particles from the capsule of the Stardust spacecraft that returned to Earth in 2006. As a result of the data obtained from this mission, scientists reported in March 2017 that extraterrestrial dust particles have been identified all over planet Earth.

Cosmic dust is the name for small amounts of matter in space, including those left over from the formation of our Solar System some 4.6 billion years ago, and new research shows that these micrometeorites are still falling on Earth billions of years later.

By one estimate, as much as 40,000 tons of cosmic dust reaches the Earth's surface every year.

So, you are made of Tinker Bell's "starstuff"!

The stuff of stardust.

> **"The cosmos is also within us. We're made of star stuff."**
>
> *Mark Twain, Life on the Mississippi*

CHAPTER 8

Galaxies

A galaxy is a circling disk of a gravitationally bound system of stars, planets, other stellar remnants, interstellar gas, dust, and dark matter.

There are two types of galaxies, the round galaxy and the spiral galaxy. There are an estimated two trillion galaxies in the universe.

As you finally approached our Milky Way galaxy on the last leg of your travels to our solar system, you will have noticed another galaxy, about twice the size of the Milky Way, and that is the Andromeda galaxy. Both are spiral galaxies.

The Milky Way contains between 200 and 400 billion other stars like our own sun and at least 100 billion planets. Some larger galaxies can contain trillions of stars.

The Milky Way is some 100,000 light years across and 2,000 light years thick and is speeding through the universe at 600 km per second.

The furthest galaxies from us are currently travelling away from us at 90% of the speed of light.

The Solar System is located within the circling disk of the Milky Way galaxy, about 26,000 light-years from the Galactic Center, on the inner edge of one of the spiral-shaped concentrations of gas and dust called the Orion Arm.

The Sun has only circled this huge galaxy some 20 times since it was formed 4.6 billion years ago.

It is thought that the Milky Way galaxy contains as many as 100 million black holes and it is believed that the centre of the Milky Way contains an enormous black hole 4 million times the mass of our Sun.

Before the true scale of the universe was realised, what is now known to be the rim of our own Milky Way was once thought to be the very edge of space.

Our neighbouring galaxy is called the Andromeda galaxy, also known as Messier31, and is approximately twice the size of the Milky Way. It is named after the Greek mythological princess Andromeda.

The Andromeda galaxy, little more than a fuzzy blur in the sky to all but the most powerful telescopes of the earliest 20th century, was considered to be a mere collection of forming stars and cosmic dust clouds. This meant it was originally called a nebula - the Great Andromeda nebula.

The center of the Andromeda galaxy is home to 26 known black hole candidates and many more have been picked out by the Chandra X-ray Observatory, launched into Earth's orbit in 1999 by NASA. Like our own Milky Way galaxy, there's also a supermassive black hole at the center of Andromeda, with two others possibly orbiting as a binary, with a mass around 140 million times that of the Sun.

Whereas most of the rest of the universe is accelerating away from our galaxy, Andromeda is blue-shifted, meaning it's moving towards us.

Both the Milky Way and Andromeda are moving towards each other at a rate of 120 kilometers (75 miles) a second, putting them on course for a galactic smash- up in around 4 billion years from now, just about the time when our own sun is due to explode and die.

This remains a possibility because although the expansion of the universe is accelerating, the galaxies themselves are not expanding.

If, for any reason, the galaxies did start to expand then, if that were to happen in the future, it may be that any collision could be avoided because all the stars, planets, etc. would have drifted so far apart by then that the two systems would simply pass through one another.

Redshift and blueshift describe how light changes as objects in space (such as stars or galaxies) move closer or farther away from us. The concept is key to charting the universe's expansion.

Visible light is a spectrum of colours, which is clear to anyone who has looked at a rainbow. When an object moves away from us, the light is shifted to the red end of the spectrum, as its wavelengths get longer. If an object moves closer, the light moves to the blue end of the spectrum, as its wavelengths get shorter.

American astronomer Edwin Hubble, who the Hubble Space Telescope is named after, was the first to describe the redshift phenomenon and tie it to an expanding universe. His observations, revealed in 1929, showed that nearly all galaxies he observed are moving away

The Andromeda galaxy was born 10 billion years ago out of the merger of many smaller protogalaxies and then, around 8 billion years ago, it ran head-on into another galaxy to form a giant that became the galaxy that we see today.

This galaxy is also a spiral galaxy although in 1998, images from the European Space Agency's Infrared Space Observatory demonstrated that the overall form of the Andromeda Galaxy may be transitioning into a ring galaxy.

It can be spotted with the naked eye and so has been known to humans for a very long time. Andromeda is accompanied by 14 dwarf galaxies. While Andromeda is the largest galaxy in the Local Cluster it may not be the most massive. The Milky Way is thought to contain more dark matter, which could make it much more massive.

The Andromeda Galaxy has a very crowded double nucleus with a massive star cluster right at its heart,

The Andromeda galaxy is so large that it would take approximately 220,000 years to cross it, if you were travelling at the speed of light.

The 2006 observations by the Spitzer Space Telescope indicates that it has an estimated one trillion stars -double that of the Milky Way. It lies three million light-years away from the Milky Way.

CHAPTER 8

Solar System

Together, our Sun and its planets, including the Earth, make up the Solar system.

Our sun is just one of billions of other suns, called stars, and these stars and their planets are held together by a force known as gravity. The Solar System began forming about 4.6 billion years ago, or about 9 billion years after the Big Bang.

The official list of planets in our star system currently now runs to eight, with Pluto being de-classified as a planet recently and gas giant Neptune now the outermost. In any event, Neptune and Pluto occasionally change places as the farthest planet from the sun because of Pluto's very elliptical orbit, so Pluto is often to be found to be closer.

Let us first look at the planets.

Mercury

The planet closest to the Sun orbiting at a distance of 36 million miles. Average temperature 750F. Has no moon.

Its orbital period around the Sun of 88 days is the shortest of all the planets in the Solar System. This is why it was named after Mercury, the Roman god of speed. As the planet also rotates very slowly, a day on Mercury is twice as long as its year.

You would imagine that Mercury would be the hottest planet in the solar system, since it is the closest planet to the Sun but in fact a planet's climate actually has more to do with atmosphere than proximity to the Sun. Venus's atmosphere is mainly carbon dioxide, making it much hotter than Mercury.

Mercury has the greatest temperature range of any planet with temperatures reaching into the hundreds of degrees, hot enough to melt tin and lead. Temperatures can then drop to -300° Fahrenheit during its long night. Mercury has practically no atmosphere.

Mercury has much more ice scattered across its north pole than previously thought, both inside craters as well as in shadowed terrain between them. It is still uncertain how water arrived on Mercury and the best guess right now is that it was delivered by water-rich comet and asteroid impacts. Another idea is that hydrogen supplied by the solar wind combined with oxygen on the planet's surface to form water.

Venus

The next planet orbiting at a distance of 67 million miles from the Sun. Hottest and brightest planet despite Mercury being much closer to the Sun. It is the second-brightest natural object in the night sky after the Moon. Venus has no tilt and therefore has no seasons of the year.

It is named after the Roman goddess of love and beauty but in fact the conditions on Venus are extremely inhospitable.

Yellow clouds of acid swirl across the planet in an atmosphere that is mostly carbon dioxide – lethal to man. Venus is hotter because of its thick atmosphere of gases and is an example of the greenhouse effect on a planetary scale. The air pressure is also 90 times greater than that of Earth and this together with the exceptionally high temperatures crushed and melted the Russian spacecraft that landed on it in 1970 in less than 2 hours.

It takes Venus 243 Earth days to rotate once on its axis, but only 223 Earth days to revolve once around the sun. For this reason, a day on Venus lasts longer than its year. Venus also rotates from west to east, meaning that the sun rises in the west and sets in the east.

Despite all this, Venus is still regarded as our sister planet - but definitely no place for a visiting Tinkerbell.

Earth

Only planet to sustain life, orbiting at a distance of 93 million miles from Sun at 30km per second. 70% covered in water. Only 10% habitable.

Mars

Fourth planet orbiting at a distance of 142 million miles from the Sun, the second-smallest planet in the Solar System, after Mercury, and about half the size of Earth.

Named after the Roman god of war, it is often referred to as the red planet. It got its name because of its rusty red color because people associated the planet's blood-red color with war. The color comes from the fact that the soil on Mars actually contains iron oxide, better known as rust. It has light red areas and darker areas and also has polar ice caps.

Mars has two very small moons, Phobos and Demos. The Martian day is about the same length as an Earth day but its year is 687 Earth days.

Mars has many interesting features on its surface, including a volcano named Mount Olympus that is the largest volcano in the entire solar system. It is almost 3 times higher than the highest peak on Earth, Mount Everest. It then has Mars's Valles Marineris canyon that is 11 times longer and four times deeper than the Grand Canyon in Arizona. The atmosphere on Mars is very thin, with temperatures on Mars averaging well below zero.

Asteroid Belt

Lying between Mars and Jupiter, and about 140 million miles across, the Asteroid Belt is home to numerous irregularly shaped bodies called asteroids or minor planets, including the dwarf planet Ceres. Ceres was originally considered a planet, but was reclassified as an asteroid in the 1850s after many other objects in similar orbits were discovered.

Jupiter

The giant stormy gas planet orbiting at a distance of 484 million miles from the Sun. The Romans named it after their god Jupiter. Jupiter is the fifth planet from the Sun and the largest in the Solar System. So big that you could fit 1300 Earths into it. It is a giant gas planet and very cold with the surface of the planet being liquid hydrogen, possibly 10,000 miles deep.

A prominent feature is the Great Red Spot, a giant storm that is known to have existed since at least the 17th century. The planet Jupiter is so large that it could fit all of the other planets of the solar system inside it.

It takes Jupiter 11.86 Earth years to orbit the sun and this huge planet has 63 moons. Jupiter takes nearly 12 Earth years to orbit the Sun and rotates exceptionally fast with one rotation taking under 10 hours.

Saturn

Another giant gas planet orbiting at a distance of 886 million miles from the Sun. Furthest planet that can be seen from Earth without a telescope. Saturn is named after the Roman god of agriculture. Saturn has a prominent ring system.

Saturn's largest moon, Titan, is the second-largest in the Solar System, is larger than the planet Mercury although less massive, and is the only moon in the Solar System to have a substantial atmosphere. However, it is the coldest place in the solar system with an average temperature of -240° C, which is completely unliveable for humans.

The Earth once had a ring like the one around Saturn. It was formed of dust and red-hot rocks, but it disappeared when the Moon was formed.

A day on Saturn lasts less than a day on Earth and its year is 29 Earth years. The planet has a stormy atmosphere made up of hydrogen and helium with wind speeds of over 1,100 miles per hour near Saturn's equator. It is these winds and the heat rising from its interior that create the bands in Saturn's outer atmosphere, visible from Earth. Saturn's south pole is a very hot spot.

Saturn has at least 62 known moons.

Uranus

Another ice giant gas planet and the coldest planet, orbiting at a distance of 1782 million miles from the Sun. Orbits lying on its side. It was discovered by William Herschel, the British astronomer, and originally called "George" after King George III, who was Herschel's sponsor, and who later funded the construction and maintenance of Herschel's 1785, 40-foot telescope.

However, common sense eventually prevailed and the planet was then renamed Uranus, the only planet whose name is derived from a figure from Greek mythology. All other planets are named after Roman gods.

To appease King George, the moons of Uranus were then named after Shakespearean characters. Before that, and because the planets all had Roman names, the moons of planets had been given names of characters associated with Greek gods.

Uranus appears a brilliant blue-green color which is due to the presence of approximately 2% methane gas in the atmosphere that is otherwise rich in hydrogen and helium. Uranus is unusual in that it rotates top to bottom or north to south due to it being tilted on its side. It therefore orbits the sun like a rolling ball. The temperature on Uranus is around -200C and its day is about 6/7 hours shorter than that of Earth. It takes 84 Earth years to travel around the sun. It also has some 13 faint rings and 27 known moons.

Neptune

The windy planet orbiting at 2,795 million miles from the sun. Neptune is four times the size of Earth, and its day lasts a little more than 16 hours. Its year is about 165 Earth years, the time taken to complete one orbit. It also appears as a light blue colour from afar but is nothing like Earth. Neptune's atmosphere is made up of helium, hydrogen, and methane. Methane absorbs red light and reflects green light, which is why Neptune appears blue.

It is a further ice giant gas planet, four times the size of Earth with 1200 mph winds that actually break the sound barrier. Neptune has at least 14 known moons in its gravitational grip, including the largest, Triton. It is the only large moon in the Solar System with a retrograde orbit, that is an orbit in the opposite direction to its planet's rotation. It is named after the Roman god of the sea. Scientists now believe that it rains diamonds on Neptune and the other ice giant Uranus as well as on the gas giants Jupiter and Saturn.

Planet 9?

Astronomers are becoming increasing certain that there is a ninth planet lurking much further out than the presently known planets.

If this is confirmed, it likely to be some ten times the size of Earth with a very wide ranging elliptical orbit.

It is thought to be located in the dark, outer reaches of the solar system, approximately twenty times farther out from the sun than Neptune.

Planet Nine's presence could explain several things, including why Kuiper Belt objects orbit in the opposite direction from everything else in the solar system and why the known planets orbit on a plane that is tilted away from the plane of the Sun's equator.

Kuiper belt

The Kuiper Belt is a ring-shaped accumulation in the Solar System beyond the planets, extending from the orbit of Neptune. It is similar to the asteroid belt, but is far larger— 20 times as wide and 20 to 200 times as massive. Like the asteroid belt, it consists mainly of some hundreds of thousands of small bodies or remnants from when the Solar System formed.

While many asteroids are composed primarily of rock and metal, most Kuiper belt objects are composed largely of frozen volatiles (termed "ices"), such as methane, ammonia and water. The Kuiper belt is home to four officially recognized dwarf planets: Pluto, Eris, Haumea and MakeMake. Some of the Solar System's moon's, such as Neptune's Triton and Saturn's Phoebe, are also thought to have originated in this region.

Pluto was relegated to the status of "dwarf planet" by the International Astronomical Union in 2006. This was due to another larger planet named Eris being discovered beyond Pluto in 2003.

The new information caused astronomers to question what really makes a planet a planet, and in the end, it was decided, based on size and location, that neither Pluto nor Eris should qualify. Eris has also been given the status of dwarf planet together with another dwarf planet, Haumea, discovered in 2004, also located in the Kuiper Belt. Pluto is smaller than our moon and Haumea only one third the mass of Pluto.

The planets are actually brighter than stars at night and they are therefore the first to be seen in the night sky.

The Oort Cloud

The Oort Cloud is a huge theoretical sphere of comets circling the Solar System some trillions of miles away from the Sun.

If the giant planet Neptune is some thirty times further out than the distance Earth is from the Sun, then the Kuiper Belt is believed to cover an area somewhere between fifty to one hundred times that distance.

Imagine then, that the Oort Cloud is thought to be as much as fifty thousand times that distance. That would be the equivalent of one light year from the Sun.

While comets orbiting the Sun and returning after less than about 200 years are known to come from the Kuiper Belt, those with a much longer-period of orbit are thought to come from the Oort Cloud.

CHAPTER 9

The Sun

The Sun is the star at the center of our Solar System which orbits the center of our galaxy, the Milky Way, at a speed of 250 km per second.

It is a nearly perfect sphere of hot plasma, with internal convective motion that generates a magnetic field via a dynamo process. It is by far the most important source of energy for life on Earth.

Temperatures in the core reach more than 27 million F (15 million C), driven by nuclear reactions.

The Sun is what is known as a G-type main-sequence star, often and imprecisely called a yellow dwarf, or G dwarf star. The Sun is in fact white, but appears yellow through the Earth's atmosphere.

Like other main-sequence stars, the Sun is converting the element hydrogen to helium in its core by means of nuclear fusion.

Each second, the Sun fuses approximately 600 million tons of hydrogen to helium, converting about 4 million tons of matter to energy.

Its diameter is about 109 times that of Earth, and its mass is about 330,000 times that of Earth, accounting for about 99.86% of the total mass of the Solar System. About three quarters of the Sun's mass consists of hydrogen. The rest is mostly helium with much smaller quantities of heavier elements, including oxygen, carbon, neon, and iron.

The Sun formed approximately 4.6 billion years ago from the gravitational collapse of matter within a region of a large molecular cloud. Most of this matter gathered in the center, whereas the rest flattened into an orbiting disk that became the Solar System.

The central mass became so hot and dense that it eventually initiated nuclear fusion in its core. It is thought that almost all stars form by this process.

The sun is one of more than 100 billion stars in the Milky Way. It orbits some 25,000 light-years from the galactic core, completing a revolution once every 250 million years or so.

The Sun is roughly middle-aged; it has not changed dramatically for more than four billion years, and will remain fairly stable for more than another five billion years.

Eventually, it will exhaust its supply of hydrogen and the core of the Sun will experience a marked increase in density and temperature while its outer layers expand to eventually become a luminous star known as a Red Giant.

In the end, it will shed its outer layers of gas, which becomes a planetary nebula, and the remaining core will collapse to become what is called a White Dwarf. Slowly, this will fade, to enter its final phase as a dim, cool theoretical object sometimes known as a Black Dwarf. A Black Dwarf star is all that is left after a White Dwarf star burns off all of its heat, but retains its mass.

It is calculated that the Sun will eventually become sufficiently large to engulf the current orbits of Mercury and Venus, and render Earth uninhabitable possibly as soon as another 1.5 billion years' time, due to the increase in temperature.

The Sun is not large enough to become a black hole when it dies.

Sunlight takes 8 minutes and 17 seconds to travel the vast distance from the sun to the earth.

The Sun is 400 times larger than the Moon and 400 times further away from Earth and this is the reason why they look to be about the same size in the sky.

Total solar eclipses can only last at most 7 minutes and 58 seconds in any given area on Earth.

The Sun is so large that over one million Earths could fit inside it. It has a radius 100 times that of the Earth. The Sun's radius is 432,700 miles and the Earth's radius is just 3,959 miles.

The enormous effect of the Sun on Earth has been recognized since prehistoric times, and the Sun has been regarded by some cultures as a deity. 'Ra' is the name of the ancient Egyptian sun god and by the 25th and 24th centuries BCE, he had become a major god in ancient Egyptian religion.

CHAPTER 10

The Moon

Originally, the Earth's moon was at a distance of only 14,000 miles from the Earth but it is slowly moving away from earth at a speed of 3.78 cm per year. It is currently 280,000 miles away. On that basis, the moon would have been twenty times larger in the sky all those years ago. That would have been some sight!

The light we see when the moon shines at night is really reflected light from the sun.

The Moon rotates at a speed of only 10 mph, compared to the Earth's rotation of 1,000 mph, but orbits the Earth at a speed of 2,300 mph. We only ever see one side of the moon because its rotation takes the same length of time as its orbit around the Earth – just under 28 days.

The moon has only one sixth the gravity of Earth. Days on the moon are very hot and the nights extremely cold. Temperatures range from -150° C to 120° C.

The four main phases of the moon, from new moon to first quarter, to full moon to fourth quarter, are due to the changing positions of the Moon and Sun relative to the Earth.

Each full moon is given a name. The Snow moon in February, the Harvest moon in September and the Hunter's moon in October.

A Blue moon is an additional full moon that appears in a subdivision of a year: either the third of four full moons in a season, or a second full moon in a month of the common calendar. The phrase has nothing to do with the actual color of the moon.

The giant-impact hypothesis, sometimes called the Big Splash, or the Big Splat, or the Theia Impact, suggests that the Moon formed out of the debris left over from a collision between Earth and an astronomical body the size of Mars, approximately 4.5 billion years ago.

The colliding body is sometimes called Theia, from the name of the mythical Greek Titan who was the mother of Selene, the goddess of the Moon.

During the course of its formation, the Earth is thought to have experienced dozens of collisions with such planet-sized bodies.

The giant-impact hypothesis is currently the favored scientific hypothesis for the formation of the Moon.

Supporting evidence includes the facts that the Moon has a relatively small iron core and has a lower density than Earth. The Earth's spin and the Moon's orbit also have similar orientations. Moon samples also indicate that the Moon's surface was once molten.

The Moon-forming collision would have been only one such "giant impact" but certainly the last and most significant impact event. Evidence also exists of similar collisions in other star systems.

An abundance of oxygen isotope in lunar rock suggests "vigorous mixing" of Theia and Earth, indicating a steep impact angle. Isotopes are variants of a particular chemical element which differ in neutron number. All isotopes of a given element have the same number of protons in each atom.

Theia's iron core would have sunk into the young Earth's core, and most of Theia's mantle accreted onto the Earth's mantle. A significant portion of the mantle material from both Theia and the Earth would have been ejected into orbit around the Earth, at just enough velocity as to allow it to remain in an orbit around the Earth, without it being thrown further out with such force as to put it into the Sun's orbit.

The material in orbit around the Earth quickly coalesced into the Moon, possibly within less than a month, but in no more than a century.

Another hypothesis attributes the formation of the Moon to the impact of a large asteroid with the Earth much later than previously thought, creating the satellite primarily from debris from Earth. In this hypothesis, the formation of the Moon occurs 60–140 million years after the formation of the Solar System.

As long ago as 1898, George Darwin, son of Charles Darwin, made the suggestion that the Earth and Moon had once been one body.

Darwin's hypothesis was that a molten Moon had been spun from the Earth because of centrifugal forces, and this became the dominant academic explanation.

Using Newtonian mechanics, he calculated that the Moon had orbited much more closely in the past and was drifting away from the Earth. This drifting was later confirmed by American and Soviet experiments, using laser ranging targets.

Like the four inner planets, the moon is rocky. It's pockmarked with craters formed by asteroid impacts millions of years ago. Because there is no weather, the craters have not eroded.

The moon has little or no atmosphere, it is practically a vacuum, so a layer of dust, or a footprint, can sit undisturbed for centuries.

Because the moon has no atmosphere, this means that the surface of the Moon is unprotected from cosmic rays, meteorites and solar winds, and has huge temperature variations. The USA stars and stripes flag, left on the Moon by Apollo 11 in 1969, has since been bleached white by radiation.

The lack of atmosphere also means no sound can be heard on the Moon, and the sky always appears black.

NASA plans to return astronauts to the moon to set up a permanent space station. Mankind may once again walk on the moon in 2019, if all goes according to plan. Both Russian and Japanese scientists are also planning to build space centers on the moon within the next 10 years.

If you look at the Moon when it is nearly full you can see the dark areas which are known as the seas. They are all given Latin names, such as Mare Serenitatis – the Sea of Serenity, or Mare Frigoris – the Sea of Cold. These are not really seas but are huge expanses of smooth dark lava.

CHAPTER 11

The Earth

So, finally, here you are living on a planet that we call "Earth" that was only formed about 4.5 billion years ago and didn't become habited by the modern human species until about 200,000 to 400,000 years ago.

Good job you didn't arrive sooner!

The most prominent scientific theory about the origin of the Earth involves a spinning cloud of dust called a solar nebula produced following the Big Bang. About five billion years ago, some of this gas and matter became our sun.

At first, it was a hot, spinning cloud of gas that also included heavier elements. As the cloud spun, it collected into a disc called a solar nebula. Our planet and others probably formed inside this disc. The center of the cloud continued to condense, eventually igniting and becoming a sun.

At first, the Earth was very hot and volcanic. A solid crust formed as the planet cooled, and impacts from asteroids and other debris caused lots of craters. As the planet continued to cool, water filled the basins that had formed in the surface, creating oceans.

Earth is the only known place where water exists in all three states: liquid, ice and vapor.

Through earthquakes, volcanic eruptions and other factors, the Earth's surface eventually reached the shape that we know today. Its mass provides the gravity that holds everything together and its surface provides a place for us to live.

The Earth is round because gravity pulls matter into a ball, although it is not perfectly round, instead being more of an "oblate spheroid" whose spin causes it to be squashed at its poles and swollen at the equator.

Earth happens to lie within the so-called "Goldilocks zone" in the Solar System where it orbits around the Sun, where temperatures are just right to maintain liquid water on its surface.

The Earth is one of four rocky planets that orbit a star that we call the Sun. The other three rocky planets are Mercury, Venus and Mars.

There are another four giant planets that orbit our sun that are made up of gas or gas and ice. They are Jupiter, Saturn, Uranus and Neptune. Some scientists are of the view that these planets still have a solid core.

Pluto is no longer considered large enough to be classed as a planet.

Earth is the only planet habitable to mankind in the Solar System and indeed, as far as we know at present, anywhere in the universe.

Even the Earth is only 10% habitable. That is because 70% of the Earth's surface is covered in water and, of the remaining 30%, 20% is either too hot around the equator or too cold at the polar regions.

Earth is composed of four main layers, starting with an inner core at the planet's centre, enveloped by the outer core, mantle and crust. Earth's interior remains active with a solid inner core, consisting mostly of iron and nickel and potentially smaller amounts of lighter elements such as sulphur and oxygen.

The outer core of the Earth is a fluid layer composed of mostly iron and nickel that lies above Earth's solid inner core and below its mantle. The mantle is made of iron and magnesium-rich silicate rocks and a convecting mantle that drives plate tectonics through a combination of pushing and spreading apart. Hot material near the Earth's core rises, and colder mantle rock sinks. Earth's outer shell is divided into several plates that glide over the mantle forming continents.

The inner core is a solid sphere made of iron and nickel metals about 759 miles in radius. There the temperature is as high as 9,800 degrees Fahrenheit (5,400 degrees Celsius). Surrounding the inner core is the outer core. This layer is about 1,400 miles thick, made of iron and nickel fluids.

The Earth's inner core spins, much like the Earth spins on its axis. The outer core spins as well, and it spins at a different rate than the inner core. creating a dynamo effect, or convections and currents within the core. This is what creates the Earth's magnetic field -- it's like a giant electromagnet.

It is also now believed that the inner and outer cores spin in opposite directions, which fits with Newton's third law of motion that states that every action has an equal and opposite reaction.

Whilst the sun creates the ozone layer, the Earth itself creates its defense against the solar wind. Without the Earth's magnetic field, ionized particles from the solar wind could strip the planet's atmosphere away. This magnetic field comes from deep inside the Earth's core.

When the solar wind reaches the Earth, it collides with the magnetic field, or magnetosphere, rather than with the atmosphere.

The poles actually change places periodically -- about 400 times in the last 330 million years.

In between the outer core and crust is the mantle, the thickest layer. This hot, viscous mixture of molten rock is about 1,800 miles thick, is solid but malleable, like plastic, and it's the source of the magma that comes from volcanoes.

The outermost layer, Earth's crust, goes about 19 miles deep on average on land. At the bottom of the ocean, the crust is thinner and extends about 3 miles from the sea floor to the top of the mantle.

Like Mars and Venus, Earth has volcanoes, mountains and valleys. Earth's lithosphere, which includes the crust (both continental and oceanic) and the upper mantle, is divided into huge plates that are constantly moving.

For example, the North American plate moves west over the Pacific Ocean basin, roughly at a rate equal to the growth of our fingernails. Earthquakes result when plates grind past one another, ride up over one another, collide to make mountains, or split and separate.

Earth's global ocean, which covers nearly 70 percent of the planet's surface, has an average depth of about 2.5 miles and contains 97 percent of Earth's water.

Almost all of Earth's volcanoes are hidden under these oceans. Hawaii's Mauna Kea volcano is taller from base to summit than Mount Everest, but most of it is underwater. Earth's longest mountain range is also underwater, at the bottom of the Arctic and Atlantic oceans. It is four times longer than the Andes, Rockies and Himalayas combined

The majority of Earth's polar regions are covered in ice, including the Antarctic ice sheet and the sea ice of the Arctic ice pack.

Earth is the densest planet in our Solar System and the largest of the four terrestrial planets.

The abundance of water on Earth's surface is a unique feature that distinguishes the "Blue Planet" from other planets in the Solar System

It was not so long ago that people believed that the Sun revolved around the Earth and was flat. They did not yet understand that the Earth was round. It was thought that explorers would sail off the edge of the World where they could see a horizon.

Earth was further believed to be the center of the universe until the 16th century when scientists first theorized that it was a moving object, comparable to the other planets in the Solar System.

The Earth's gravity interacts with other objects in space, especially the Sun and the Moon. The gravitational attraction between Earth and the Moon causes tides on Earth. The moon is the Earth's only natural satellite.

During one orbit around the Sun, the Earth rotates about its axis about 365 times; thus, an Earth year is about 365 days long. Each rotation is therefore one day on Earth. The Earth orbits the Sun at an average distance of about 93 million miles.

The Earth spins on its axis at about 900 miles per hour.

The axial tilt of the Earth is approximately 23° with the axis of its orbital plane always pointing towards the Celestial Poles. Due to Earth's axial tilt, the amount of sunlight reaching any given point on the surface varies over the course of the year.

This causes the seasonal change in climate, with summer in the Northern Hemisphere occurring when the Tropic of Cancer is facing the Sun, and winter taking place when the Tropic of Capricorn in the Southern Hemisphere faces the Sun.

During the summer, the daylight lasts longer, and the Sun climbs higher in the sky. In winter, the climate becomes cooler and the daylight shorter.

In the Northern Hemisphere, the winter solstice, the shortest daylight of the year, currently occurs around 21 December.

The summer solstice or longest day is around 21 June.

In the Southern Hemisphere, the situation is reversed, with the summer and winter solstices exchanged and the spring and autumnal equinox dates swapped.

The Earth's atmosphere is mostly composed of nitrogen, (78 per cent). Oxygen, produced by plants, then makes up just 21 percent of the air we breathe. Carbon dioxide, argon, ozone, water vapor and other gasses make up a tiny portion of it, as little as 1 percent. Some scientists believe that our atmosphere had little to no oxygen before plants evolved and started releasing it.

The atmosphere has five primary layers, from highest to lowest:-

- *Exosphere: 700 to 10,000 km (440 to 6,200 miles)*
- *Thermosphere: 80 to 700 km (50 to 440 miles)*
- *Mesosphere: 50 to 80 km (31 to 50 miles)*
- *Stratosphere: 12 to 50 km (7 to 31 miles)*
- *Troposphere: 0 to 12 km (0 to 7 miles)*

We live in the lowest layer of the atmosphere known as the troposphere, which contains three quarters of the atmosphere's mass and is constantly in motion, creating the weather. Sunlight heats the planet's surface, causing warm air to rise. This air ultimately expands and cools as air pressure decreases, and because this cool air is denser than its surroundings, it then sinks, only to get warmed by the Earth once again.

Some 30 miles above the Earth's surface, and above the troposphere is the stratosphere. The still air of the stratosphere contains the ozone layer, which was created when ultraviolet light caused trios of oxygen atoms to bind together into ozone molecules. Ozone prevents most of the sun's harmful ultraviolet radiation from reaching Earth's surface.

Roughly 100 miles above Earth, in the exosphere, the air is so thin that satellites can shoot around with little resistance. Still, traces of atmosphere can be found as high as 370 miles above the surface. The aurora borealis and aurora australis sometimes occur in the lower part of the exosphere, where they overlap into the thermosphere.

Water vapor, carbon dioxide and other gases in the atmosphere trap heat from the sun, warming Earth. Without this so-called "greenhouse effect"," Earth would probably be too cold for life to exist, although a runaway greenhouse effect would lead to the horrendous conditions now present on Venus.

Earth-orbiting satellites have shown that the upper atmosphere actually expands during the day and contracts at night due to heating and cooling.

Earth is the only planet in the universe known to possess life. There are several million known species of life, ranging from the bottom of the deepest ocean to a few miles into the atmosphere, and scientists think far more remain to be discovered. Scientists figure there are between 5 million and 100 million species on Earth, but science has only identified about 2 million of them.

The name Earth is at least 1,000 years old. All of the planets, except for Earth, were named after Greek and Roman gods and goddesses. However, the name Earth is an English/German word, which simply means the ground.

So, here you are on a planet that is literally just one of billions of billions of billions in the Universe, sitting in a Solar System that is just a small part of our Milky Way galaxy and which, in turn, is just one of trillions of other galaxies in the Universe.

You look up at the stars at night and you now know that what you thought were simply stars are in fact probably whole galaxies, far, far away.

The stars or galaxies do not seem to move as you watch them although you notice that the brighter ones that are the first to show in the night sky do indeed move around. This is because they are the planets, the wanderers, orbiting the Sun like us and the light coming from them is simply the reflected light of the Sun.

Although you have no sense of movement you are actually on a planet that is spinning at a speed of 900 mph and orbiting the Sun at a very fast 66,000 miles an hour. The Sun is then orbiting the Milky Way galaxy at a speed of 515,000 mph and then the Milky Way galaxy itself is travelling through the Universe at nearly 1,350,000 miles an hour.

I told you that your orienteering skills were phenomenal!

So, if you want to wish upon a star there is a very good chance you will miss unless, of course, you wish upon our own star, the Sun.

CHAPTER 12

The Future

So, what of the future?

This really requires a two-part answer.

One part to deal with the question of the future of the universe and the second to deal with the question of your future.

As far as science is concerned you can see that it is well equipped and prepared to continue its exploration of the universe. At present, it is impossible to say just how things will develop from this point on because there are still so many imponderables that we cannot be certain of at present.

As far as the Earth is concerned, of the different possibilities, one of the first changes might be a shift in the axial tilt which will reverse, causing the seasons to reverse with the northern hemisphere winter starting in June.

Then, after about 500,000 years, nuclear waste will have become safe

In another 4 million years, the Straits of Gibraltar will close up again and the Mediterranean Sea will have dried up

After 15 million years, the East African rift will have widened and the Somali plate will have broken away.

After 40 million years, Antarctica will have become ice-free and after 50 million years Africa will collide with Eurasia.

In 140 million years from now, Africa and America will start moving back together

After 240 million years, the Solar System will have completed one orbit around the Milky Way's Galactic Core and a new supercontinent will have formed only to disappear again after about 400 million years.

After 900 million years, the Earth's oceans will evaporate.

In 1.5 billion years the Earth will become uninhabitable due to the high temperature.

In 2.3 billion years from now, Earth's outer core will freeze and after 2.8 billion years the average temperatures on Earth will reach 420K.

In 3.5 billion years the Earth's surface conditions now become similar to that of present day Venus and after 4 million years our Milky Way galaxy will start colliding with Andromeda galaxy.

After 5 billion years the Sun runs out of oxygen and starts the process of evolving into a red giant.

After 7.9 billion years the Sun reaches its largest size and Mercury and Venus vaporize, possibly Earth as well and after 8 billion years the Sun transforms into a white dwarf.

Not a lot of positives there. However, against that we know that for starters, we are not going to live that long anyway. Also, nearly a fourth of sun-like stars observed by Kepler have potentially habitable Earth-size planets and NASA thinks we will find signs of alien life in as little as 10 years—and "definitive evidence" by about 20 years. Now that's exciting!

Add to this the fact that private enterprise is already well on the way to promoting space travel in the near future and there is a lot to look forward to.

As to the Universe as a whole, having started with the Big Bang, one possibility is that it will end with the Big Crunch. This supposes that the continued expansion of the universe will eventually stop and reverse, causing the universe to collapse to a point when this may even trigger the whole Big Bang all over again.

This depends on the closed universe theory but there is still disagreement as to whether the universe is open or closed.

Some more recent theories assume the universe may have a significant amount of dark energy, whose repulsive force may be sufficient to cause the expansion of the universe to continue forever.

If the universe is open then it will continue to expand forever, even without the force of dark energy which will simply speed up the expansion.

With an open universe if the expansion of the universe were to continue forever then another theory is that everything in the universe will become so far apart that the universe will cool as it expands, eventually becoming too cold to sustain life. For this reason, this future scenario is popularly called heat death or the Big Freeze.

A further possibility in an open universe is the Big Rip - in which all the matter of the universe, from stars and galaxies to atoms and subatomic particles, and even spacetime itself, is progressively torn apart by the expansion of the universe at an unknown time in the future.

There is currently a strong overall consensus among cosmologists that the universe is flat and will continue to expand forever.

In the absence of dark energy, a flat universe expands forever but at a continually decelerating rate.

With dark energy, the expansion rate of the universe initially slowed down, due to the effect of gravity, but is now increasing. The ultimate fate of the flat universe is the same as an open universe.

There is then the Big Bounce. The concept of the Big Bounce envisions the Big Bang as the beginning of a period of expansion that followed a period of contraction. In this view, one could talk of a Big Crunch followed by another Big Bang, or more simply, a Big Bounce.

This suggests that we could be living at any point in an infinite repetition of the universe, or conversely the current universe could be the very first in the sequence.

We have seen that the strongest evidence suggests that the Earth will become uninhabitable to mankind in something like 1.5 billion years' time due to the increasing heat from the sun.

Ultimately, it seems that the eventual fate of the universe may well depend on Dark Energy and what influence this eventually exerts.

Another school of thought, to perhaps concentrate your mind, is to the effect that Homo Sapiens will not survive another 1,000 years but rather will become superseded by some sort of race of superhumans.

However, if nothing else you can be sure that the journey into the future will continue to be just as exciting and mysterious as it has been in the past.

That is some of the possible events affecting the future of the universe.

But what of your future?

> **"The best way to predict the future is to create it."**
>
> Abraham Lincoln, 16th President of the United States

As far as you are concerned, well, you have seen just how special you are.

Clearly, you have also seen how very, very fortunate you have been in getting to where you are now - but it is your good fortune. You do not owe your good fortune to anyone else.

Whatever your position in the world you owe it to yourself to make the most of your life going forward.

Hopefully, you will be in a position to achieve most if not all of your life's goals.

However, it is a sad fact that many of the world's population are not in such a fortunate position, this depending largely on where you live in the world, the influence of your country's politics and religions and even what sex you are.

The fact is that if you are in any way dependent on others for your emotional health, lifestyle, or if you need someone else to tell you what to do, you are not in control of your life.

If that is so, you should now look to work towards improving your present situation, in whatever way you feel is necessary, so as to take complete control of your life, whilst always being sure to keep yourself out of harm's way.

You should think about what you value most so that you can concentrate on getting these values realised. Think about what and who is important to you — is it freedom, happiness, equality, your family?

If you have a job or a relationship or anything else imposed on you that lacks respect, then it will help you to realize that you are not prepared to accept this situation lying down and that you are actively looking to do something about it.

It also sends a message to those around you that you are not someone to try and manipulate or walk over and if you are able to achieve this then it will bring you less frustration as well.

You will need to have wisdom. Wisdom encourages you to gain knowledge and experience so that you can use information to your advantage.

You will need to have courage. Having courage means you draw on your strength and willpower to accomplish what you need or want to do, despite some form of adversity.

Do not live for others any more than you would expect others to live for you.

If you are suffering any sort of abuse, whether physical or emotional, including sexual, then if you cannot change an abusive relationship you should seriously consider leaving it.

It is very difficult for anyone to break up a relationship and especially hard if they are being abused but you owe it to yourself.

Remember, abuse is not normal in a relationship, and it's not OK.

You may feel that you face problems that are truly beyond your capabilities. In such cases you should seek outside help where possible.

Don't let these 13.8 billion years go to waste!

It is perhaps useful to make some mention of religion here.

> *Life is but a momentary glimpse of the wonders of this amazing Universe, and it is sad to see so many dreaming it away on spiritual fantasy.*
>
> -Andy Mulcahy

Science tells us that there is no actual evidence of the existence of any god in the whole 13.8 billion years that the universe has been evolving. Over 500 years of science tells us that any God is not only unlikely – it is irrational.

The earliest evidence of religious practice dates back as far as about 40,000 BCE. There are an estimated 10,000 distinct religions worldwide and there are 5,000 different gods. That alone tells us that gods are man-made.

> *"People ask me if a god created the Universe, I tell them the question itself makes no sense.*
>
> *Time didn't exist before the Big Bang, so there is no time for God to make the universe in.*
>
> *It's like asking for directions to the edge of the Earth. The Earth is a sphere. It does not have an edge, so looking for it is a futile exercise."*
>
> — Stephen Hawking, British Cosmologist, Theoretical Physicist

If we happen to belong to one religion or another it is almost certainly because we just happened to be born in a particular part of the world, not because we actually believe in that religion or because we have chosen one religion over another.

Now ignorance is not a crime and you have to feel for those born hundreds or thousands of years ago who would have been absolutely terrified for their lives every single day and night by so many events such as volcanic eruptions, earthquakes, tsunamis, even thunder and lightning. As well as a whole host of wild animals and hostile tribes, there would have been droughts, famine and floods.

There was no knowledge of diseases and illness and no medicines, and having no control over events meant that it was convenient always to look around for the possible cause and to lay the blame anywhere and everywhere.

If you were fortunate enough to survive childbirth you would then be faced with a very short expectation of life, often as little as 25 years and you would die of all sorts of complaints that would be easily curable today, thanks to science.

Small wonder then that early man would have invented witches and devils and angry gods as the cause of their troubles and they would have been very keen to appease the gods in whatever way occurred to them, even to the extent of offering animal or even human sacrifices to try to ward off these perils.

We can no longer plead such ignorance. Hopefully no one these days is going to rush out and offer a sacrifice because they hear thunder or see lightning. We now know much better and the world needs to continue to wake up from its long nightmare of religious beliefs, dogmas, myths and legends and adjust itself to the modern day more quickly.

Thanks to science we can at last let go of all of the ignorance and superstitions that have caused such havoc for centuries now and concentrate on appreciating just who we are, what we are and where we are. We cannot go on living in the dark ages.

Religious dogmas, myths and superstitions just build hatred and separation between people.

We have to face facts when the evidence is beyond doubt.

> *"Facts do not cease to exist because they are ignored"*
>
> *Aldous Huxley, British Philosopher*

And just think. If god were real would he be so difficult to find? We know that there are approximately 6,800 languages in the whole world. Would he be speaking to just a few minority groups in remote parts of the world rather than to major language speakers such as the Chinese, English and Spanish?

If there is a god he will surely make himself known to everyone at the same time and until he does that then we can be better served by continuing to gain knowledge and improving ourselves.

It is estimated that well over 100 billion people have died on our planet so far since humans evolved. No one came to their aid over the past millions of years so what is the likelihood that someone will do so now in the immediate future?

What good did it do them to spend their lives beholding to various mythological gods?

Even now, with all the advances of modern medicine, over 2.7 million babies still die every year in their first month of life and a similar number are stillborn.

Some 21,000 children die every day around the world. That is equivalent to just under 7.6 million children dying every year.

In the first 10 years of this 3rd millennium, some 92 million children have already died.

You have to wonder just how many unanswered prayers were said for those children?

They weren't even given the chance of trying to guess which religion might just be true.

Does this sound even remotely like a grand design made with you in mind? Should this be the case in a world controlled by a god?

If and when there ever is a god then these are questions that will need to asked.

So, don't be afraid that your life will end, be afraid that it will never begin.

Just remember, it is the stars that die to save humankind – no one else!

> *Our honorable and worthy ancestors knew that the world was flat, motionless, and the center of the universe.*
>
> *They knew the human body could not withstand the forces of travelling faster than 19 mph.*
>
> *They knew that the way to salvation was exorcising witches and slaying non-believers.*
>
> *They knew that it was a mortal sin to marry someone of a different skin color.*
>
> *- Jonathan Lockwood Huie*

CHAPTER 13

Quiz

(answers after the Glossary)

1. How many miles are there in one light year?
2. What percentage of the Universe is taken up with ordinary matter?
3. How long ago did the dinosaurs become extinct?
4. What were our appendix originally used for?
5. Approximately how many cells does the human body contain?
6. Approximately how many new stars are born every day?
7. How far away is the Andromeda galaxy from the Milky Way?
8. On how many planets is it thought to rain diamonds and can you name them?

9. Which planet rotates from top to bottom, or North to South?

10. How far from the Galactic Core does the Sun orbit?

11. Why does the moon appear to be the same size as the Sun?

12. What percentage of the Earth is habitable by man?

13. How soon does NASA think we will find evidence of alien life?

CHAPTER 14

Timeline of Universe

- **13.8 Billion years ago (BYA)** - Big Bang –The beginning of time and space
- 13.5 BYA - first black holes
- 13.2 BYA - first galaxies form
- 12.6 BYA - Milky Way forms
- 11.5 BYA - oldest known exoplanet lying within habitable zone - An exoplanet is a planet outside of our solar system that orbits another star.
- 11.4 BYA - Milky Way's cosmic halo forms
- 9.7 BYA - universe reaches one third of its present diameter
- 9.0 BYA - oldest known evidence for dark energy
- 8.8 BYA - Milky Way's thin disk starts to form
- 8.0 BYA - Milky Way star formation rate starts to decline

- 7.1 BYA - universe cools to below 5 K
- 6.9 BYA - half age of universe
- 6.5 BYA - Milky Way starts spiraling
- 6.0 BYA - dark energy overtakes gravity - universe expansion rate starts to accelerate
- 5.5 BYA - Milky Way becomes a spiral galaxy
- 4.57 BYA - Sun forms
- 4.56 BYA – primitive Earth forms – planets asteroids and comets form from the Sun's protoplanetary disk
- 4.53 BYA - Moon forms
- 4.4 BYA - oceans form on earth from water brought by asteroids
- 4.2 BYA - life emerges on earth – early estimate
- 3.9 BYA - life emerges on earth – late estimate
- 3.6 BYA - Valbaara supercontinent emerges
- 3.25 BYA - Barberton event – one of largest impacts ever in South Africa
- 3.2 BYA - Ur supercontinent forms
- 3.1 BYA - Kenorland supercontinent forms
- 2.8 BYA - Valbaara supercontinent breaks up
- 2.5 BYA - movement of tectonic plates commences

- 2.4 BYA - Suavjarvi impact event in Russia – oldest still recognizable crater - Huronian glaciation starts – first snowball earth period

- 2.22 BYA - atmospheric oxygen level surpasses one percent

- 2.2 BYA - ozone layer forms

- 2.15 BYA - Huronian glaciation ends - It is the oldest and longest ice age, occurring at a time when, in a biological sense, only simple, unicellular life existed on Earth.

- 2.1 BYA - earliest multicellular life

- 2.02 BYA - Vredefort impact event in South Africa – one of the largest impacts ever

- 1.9 BYA - atmospheric oxygen levels surpass 15 percent

- 1.85 BYA - Sudbury impact event in Ontario, Canada

- 1.8 BYA - supercontinent Columbia forms

- 1.5 BYA - supercontinent Columbia breaks up

- 1.2 BYA - supercontinent Rodinia forms

- **770-740 Million years ago (M)**– Kaigas glaciation – snowball earth period

- 750 M - supercontinent Rodinia breaks apart

- 720-760 M - Sturtian galacian -snowball earth period

- 650 – 635 M - Marinoan glaciation – snowball earth

- 540 – 520 M - Cambrian explosion – exponential diversification of life
- 445 M - Ordovician extinction event – 60% of all species go extinct
- 420 M - first air-breathing animals
- 395 M - first tetrapods - four limbed animals
- 380 M - first tree like plants
- 370 M - late Devonian extinction – 70% of all species go extinct
- 320 M - first reptiles
- 300 M - Pangaea forms – last supercontinent
- 251 M - Permian extinction event – largest extinction event in history – 90-96% of all species go extinct
- 231 M - first dinosaurs
- 225 M - first mammals
- 201 M- Triassic extinction event – 75% of all species go extinct
- 180M - Pangaea splits into Laurasia and Gondwana
- 176M - first stegosaurs
- 155M - first birds
- 145M - Madagaskar splits from Africa
- 130M - Laurasia and Gondwana start to drift apart first flowering plants

- 106 M - Spinosaurus evolves – largest of all carnivorous dinosaurs
- 100 M - first bees
- 90 M - Indian subcontinent splits from Gondwana
- 80 M - Australia splits from Antartica
- 68 M - Tyrannosaurus rex evolves
- 66 M - Cretaceous extinction event in Mexican basin – 75% of all species including non-avian dinosaurs go extinct
- 60 M - first primates evolve
- 55 M - first modern birds
- 50 M - Indian subcontinent collides with Asia – Himalayas start to rise
- 40 M - Antarctic ice cap begins to grow
- 35 M - grasslands become widespread
- 30 M - South America separates from Antarctica
- 26 M - first elephants
- 18 M - divergence of great apes and lesser apes
- 15 M - first bovids
- 11 M - first large horses
- 6.0 M - last common ancestor humans and chimpanzees

- 5.96 – 5.33 M - Messinian Salinity crisis – Strait of Gibraltar closes tight and Mediterranean Sea dries up

- 4.0 M - Australopithicus evolves

- 3.0 M - great American interchange - North and south America join

- 2.5 M - current Ice Age begins

- 2.2 M - first members of the genus Homo evolve

- 1.5 M - first controlled use of fire by homo erectus

- 780,000 years ago - last reversal of the Earth's magnetic field

- 500,000 - earliest human engravings

- 300,000 - earliest instances of burial

- 250,000 - Neanderthals evolve

- 200-350,00 - anatomically modern humans appear in Africa

- 100,000 - humans begin migrating out of Africa

- 70,000 Toba super-volcanic eruption – human population falls to 10,000

- 38,000 - the Neanderthals go extinct

- 36,000 – the first domesticated dogs

- **6,000 - to present time** - human civilization span of recorded history

CHAPTER 15

Glossary

A

Accretion disk – an increase of material, mainly gases, around a celestial object such as a black hole

Andromeda galaxy – the nearest galaxy to our own Milky Way

Asteroid – irregular shaped bodies in the Solar System, mainly composed of mineral and rock, orbiting the sun and located roughly between the orbits of the planets Mars and Jupiter

Atmosphere – gases around the Earth held under the force of gravity

Atom – the smallest quantity of an element that can take part in a chemical reaction. Made up of protons, neutrons and electrons, it is the building block of matter

Aurora australis – bright bands of glowing light that appear in the night sky near the South Pole. Also called the southern lights

Aurora borealis – bright bands of glowing light that appear in the night sky near the North Pole. Also called the northern lights

---

B

Baryon – class of elementary particle that decays into a set of particles that include a proton

Big Bang – the name commonly given to the start of the universe

Big Bounce – the name given to the start of the universe, assuming that it restarts after previously ended in a Big Crunch

Big Crunch – the name given to the possible collapse of the universe after expansion ceases

Big Splash – the name given to the giant-impact hypothesis, where the Earth is thought to have been in collision with another Mars size body and which resulted in the formation of the moon

Binary star – a star that comprises two stars orbiting each other

Black Hole – an extremely dense area of matter formed by the death and gravitational collapse of a star. So dense that not even light can escape

C

Comet - A small, frozen mass of dust and gas revolving around the sun.

Constellation – a group of stars in the sky, often given the names of animals, objects or people

Corona – the outer atmosphere of our Sun or other stars

Cosmos – the universe considered as an ordered system

D

Dark Energy – hypothetical force of negative energy that is responsible for the expansion and acceleration of the universe

Dark Matter – matter that takes up a substantial part of the universe but that cannot be seen but can be presumed from the gravitational effect on other bodies

Dwarf planet – small planets whose orbits around the Sun is shared by other smaller objects

E

Electron – negatively charged elementary particle found outside of the nucleus of an atom

Equator – the imaginary line around the Earth or other body that is equidistant between the north and south poles

Event horizon – the rim or area around a black hole

Exoplanet – a planet outside our Solar System that orbits a star

G

Galaxy – a star system held in place by gravitational attraction

Gamma rays – also called gamma radiation, they come from solar flares and have lots of energy

Gravitational lens - a galaxy or group of galaxies that bend light rays from a background object

Gravity – a natural phenomenon whereby all things with mass gravitate towards each other

Greenhouse gases – gases, such as carbon dioxide, ethane and nitrous oxide that are held in the atmosphere that trap heat from the Sun

K

Kuiper Belt – a ring of icy objects, including the dwarf planet Pluto, beyond the orbit of Neptune

L

Large Hadron Collider – the World's largest underground particle accelerator

Light year - A light year is a way of measuring distance where those distances are gigantic. A light year is 6 trillion or 5,865,696,000,000 miles and is the distance that light would travel in a year

M

Magnetar – see neutron star

Magnetic field – the area around a magnet where that magnetic force is active. The Earth has a magnetic field that shields it from Space weather

Mass – the amount of matter of an object

Matter – what everything is made of

Meteor – a streak of light from a meteoroid entering a planet's atmosphere heating up due to friction

Meteorite – a meteoroid that collides with the surface of a planet

Meteoroid – essentially, a small asteroid

Milky Way – a spiral galaxy that contains our Solar System

Molecule – the smallest physical unit of an element or compound, consisting of one or more like atoms in an element and two or more different atoms in a compound

Moon – Earth's only permanent natural satellite

N

Nebula – a cloud of gas or dust found between stars

Neutron – an elementary subatomic particle with a mass slightly larger than a proton

Neutron star – the very dense collapsed core of a massive dying star

Nucleus – the central part of an atom consisting of protons and neutrons

O

Oort cloud – an orbiting cloud of icy material and rocks on the very outer edge of the Milky Way

Orbit – the curved path of a planet, satellite or spacecraft as it circles around another object

Ozone layer – the part of the atmosphere that absorbs lots of the Sun's ultraviolet radiation

P

Particle - a small object to which can be attributed several properties such as volume or mass

Planet – an astronomical body orbiting a Star like our Sun. Planets shine by reflected light; stars shine by producing their own light

Plate tectonics – the outer shell of the Earth is divided into several plates that glide over the inner layer above the mantle that in turn sits on the core

Polar cap – the area at the poles of either the Arctic or Antarctic circles

Prism – transparent polygonal solid that can have the effect of bending light

Proton – a positively charged elementary subatomic particle

Pulsar – a rapidly rotating neutron star releasing short pulses of radio waves

Q

Quantum mechanics – the science of the very small. The study of the atom and elementary subatomic particles

Quasar – a supermassive black hole emitting immense amounts of energy in the form of light from its event horizon perimeter

S

Sagittarius A – a super massive black hole at the center of the Milky Way with a mass of 4.3 million times that of the Sun

Solar flare – a burst of energy from the Sun emitting gases, radiation waves and magnetic storms that we observe on Earth in the northern and southern lights

Solar system – a gravitationally bound system comprising the Sun and orbiting planets, including the Earth

Solar wind – the constant stream of particles and energy emitted by the Sun

Space time – a concept that recognizes the union of space and time as suggested by Albert Einstein in his theory of relativity

Speed of light – light is the fastest thing in the Universe, travelling at 186,282 miles per second

Star – a luminous sphere of plasma, such as the Sun, emitting huge amounts of energy from its nuclear cauldron

Stellar – of or relating to a star or stars

Sub-atomic - of or relating to particles that are smaller than an atom

Sun – the star at the centre of our Solar System

Supernova – the explosion of a giant star that has reached the end of its life

---

T

Tectonics – movements of the Earth's crust that creates mountain ranges, volcanoes, earthquakes and tsunamis

---

U

Universe – the total of all existing matter, energy and space

---

V

Vacuum – an empty space containing no matter

CHAPTER 15

Answer to quiz questions

1. There are 6 trillion miles in one light year

2. Ordinary matter comprises just 4% of the Universe

3. Dinosaurs became extinct 66 million years ago – long, long before man appeared

4. Our appendix was originally used to produce cellulose to break down plants but is now obsolete

5. The human body contains about 100 trillion cells

6. Approximately 275 million stars are born every day

7. The Andromeda and Milky Way galaxies are about three million light years apart

8. Four –Jupiter Saturn Uranus and Neptune

9. Uranus

10. The Sun orbits about 25,000 light years from the Galactic Core
11. Although the Sun is 40 times larger than the Moon it is also 40 times further away from the Earth
12. Only 10% of the Earth is habitable to man
13. NASA believes that we will find evidence of alien life in just 10 years from now.

*It seems to me that the natural world is the greatest source of excitement;
the greatest source of visual beauty;
the greatest source of intellectual interest.*

It is the greatest source of so much in life that makes life worth living.

—DAVID ATTENBOROUGH

www.ingramcontent.com/pod-product-compliance
Lightning Source LLC
Chambersburg PA
CBHW070302230526
45470CB00002B/678